啖啖肉

Meat & Poultry

Home-made Cafe

自煮飯堂

U0106781

目錄

Contents

雞全翼
Whole Chicken Wings

包括雞槌、雞中翼及雞翼尖，各有多變的烹調方法，深受男女老幼喜愛。市面出售的急凍雞全翼多來自美國及巴西。

Each of them is made up of a drumette, a winglet (mid-joint) and the tip. Each part can be cooked separately in various ways and they are loved by everyone. Most frozen whole chicken wings available in the market are either from the U.S. or Brazil.

小雞槌
Chicken Drumette

質感嫩實，常用作煎炸。由於小雞槌肉厚，醃味或烹煮需時較長。

With a firm and tender texture, chicken drumettes are usually shallow- or deep-fried. As the flesh on the drumettes is thick, make sure you allow more time for marinating or cooking.

雞中翼
Mid-joint Chicken Wings

肉質嫩滑可口，加上肥厚的外皮，汁液豐盈。除原隻烹調外，亦可去掉雞骨，釀入餡料，同樣滋味。

Their incredibly tender and succulent meat is wrapped in juicy and plump skin. Besides cooking them as they are, you may also bone and stuff them for a variation.

雞翼尖
Chicken Wing Tips

皮多肉少，多作涼菜或佐酒小吃。烹調前先飛水，可減少油膩。

There is more skin than meat on the wing tips. They are usually made into cold appetizers or snacks to go with alcoholic drinks. To make the wing tips less greasy, you may blanch them first before use.

豬扒
Pork Chops

宜選骨肉均勻及帶少許脂肪的，緊記先用刀背拍鬆豬扒。煎豬扒時，先用大火封着肉汁，再以小火煎熟，勿常翻轉及煎得太久。

For tenderness and succulence, pick those bone-in pork chops with some fat on. Make sure you tenderize the pork chops with the back of a knife. Fry them over high heat to seal in the juice first. Then turn to low heat to cook them through. Resist the temptation to flip them often or fry them for too long.

牛扒
Steak

分為西冷扒、T骨扒及肉眼扒，肉質各異。宜用生粉醃製，令肉質嫩滑。

Steak refers to different cuts of beef, such as sirloin, T-bone and rib eye among other less tender cuts. They have different textures and are in different degree of beef flavour. It's advisable to add a little caltrop starch to the marinade, which makes the beef more juicy and tender.

凍肉快速解凍法 Do and Don't
Do and Don't of Thawing Frozen Meat

Do

- 放在室溫數小時，或將凍肉放於冰箱下層解凍。
- 若想縮短解凍時間，建議將凍肉包好，放入冷水或鹽水中浸泡，既可縮短解凍時間，又不會滋生細菌。
- 將肉切成細塊，增加肉類接觸氣溫的面積，加速解凍。
- 利用微波爐以分階段形式，逐分鐘解凍。

- Leave the meat at room temperature for a few hours. Or put it on the lower shelf of a refrigerator overnight.
- To shorten the thawing time, you may wrap the meat well in cling wrap first. Then soak it in cold water or salted water. You can speed up the thawing process without encouraging bacterial growth this way.
- Cut the frozen meat into pieces to increase the surface area that gets in touch with the air. The meat will be thawed more quickly.
- Thaw it in a microwave oven in stages. Thaw for one minute each time and check on the meat texture after each operation.

Don't

切忌用熱水浸泡凍肉，會失去部份蛋白質及香味。

Do not soak the meat in hot water. You're partly cooking the meat and it is more likely to turn stale. Besides, you lose part of the protein and meat flavour this way.

薑
Ginger

味香辛，烹調前在熱水中加入薑片，再放入肉類飛水，有效去除冷藏味；又或在醃味時加入薑片，但肉類會帶辛辣味。

Ginger is spicy and aromatic. Before you cook, blanch it in a pot of water with a slice of ginger to remove the unpleasant smell sometimes present in frozen meat. Or, you may add a slice of ginger to the marinade for similar effect. However, ginger also adds some spiciness to your dish.

4大伙伴
Top 4 Companions to Frozen Meat
有效去除凍肉冷藏味
to Taste as Good as Fresh Meat

葱
Spring Onion

一般用薑葱飛水辟味，或烹調時加入葱段，增添清香之氣，又能辟味。

Most people put ginger and spring onion in the water before blanching meat to remove unpleasant smell. You may also put in spring onion in the cooking process to add an extra oniony taste to the dish.

蒜
Garlic

蒜香味強烈及持久，能辟除羊肉、牛肉等羶味。蒜頭經烹煮後，沒有濃烈的辛辣味，還帶一陣甜味，為菜餚帶來獨特的香氣。

With its pungent and lasting taste, garlic is commonly used in marinade to cover the goaty or gamey taste of lamb or beef. Garlic doesn't taste pungent and spicy after being cooked. In fact it is quite sweet while adding a characteristic fragrance.

紹酒
Shaoxing Wine

紹酒醇香，多在炒煮時加入，增添香氣之餘，能辟除肉類的冷藏味，令人增加食欲，提升肉類鮮香之氣。

Owing to its mellow taste and flowery nose, Shaoxing wine is commonly "sizzled" on the hot wok in stir-fries. It also covers up the unpleasant smell of frozen food. Shaoxing wine whets the appetite and heightens the meaty taste of dishes.

香橙豬扒
Pork Chops in Orange Sauce

材料 Ingredients

豬扒 4 塊，橙 2 個，橙皮絲數條

4 pieces pork chop, 2 oranges, shredded orange zest

醃料 Marinade

鹽半茶匙，生抽 1 湯匙，糖 1 茶匙，
紹酒 1 茶匙，生粉 1 湯匙，蛋液 2 湯匙

1/2 tsp salt, 1 tbsp light soy sauce, 1 tsp sugar, 1 tsp Shaoxing wine, 1 tbsp caltrop starch, 2 tbsps egg wash

獻汁 Sauce

橙汁 3/4 杯，醋半杯，片糖半塊，生粉 1 茶匙

3/4 cup orange juice, 1/2 cup vinegar, half slab brown sugar, 1 tsp caltrop starch

做法 Method

1. 豬扒洗淨，用刀背剁鬆，拌入醃料醃 30 分鐘。
2. 橙皮切絲；橙榨汁備用。
3. 燒熱油，下豬扒炸透至熟透及金黃色。
4. 煮滾獻汁，下橙皮絲，豬扒回鑊煮片刻即成。

1. Rinse pork chops. Pat lightly with the flat of the knife. Mix with marinade for 30 minutes.
2. Shred the orange and squeeze the orange juice.
3. Heat the oil. Deep fry the pork chops until done and golden.
4. Bring the sauce to the boil. Add the orange shreds and return the pork chops. Stir fry for a while. Serve.

豬扒烹調時要拍鬆，令肉質更鬆軟。

Make sure you tenderize the pork chops by tapping them with the back of a knife (or a meat hammer) before use. Otherwise, they might be tough and chewy.

芝 士 釀 豬 扒

Stuffed Pork Chops
with Cheese

豬扒 8 塊，芝士 8 片，火腿 8 片，麵粉適量

8 pieces pork chop, 8 slices cheese, 8 slices ham, flour

鹽 1 茶匙，黑胡椒碎少許

1 tsp salt, crushed black pepper

1. 豬扒洗淨，瀝乾水分，起雙飛（即橫切一半，但不切斷），用醃料醃 30 分鐘。
2. 豬扒拍上適量麵粉，夾上 1 片芝士和 1 片火腿。
3. 燒熱油，下豬扒煎至熟，盛起即可。

1. Wash and wipe dry pork chops. Butterfly the pork chops. Marinate pork chops for 30 minutes.
2. Coat pork chops with some flour. Then stuff with 1 slice of cheese and 1 slice of ham.
3. Heat the oil. Fry pork chops until done. Remove and ready to serve.

紅 酒 黑 醋 燴 豬 扒

Braised Pork Chops with Red Wine and Balsamic Vinegar

材料 Ingredients

豬扒 4 塊，洋蔥 1 小個（切粗絲），
番茄 1 至 2 個（切塊），蒜茸 1 茶匙

4 pieces pork chop, 1 small onion (shredded thickly), 1-2 tomatoes (cut into pieces), 1 tsp grated garlic

醃料 Marinade

蒜茸 1 茶匙，鹽半茶匙，黑椒粉少許，糖半湯匙，
油 1 茶匙，水 1 湯匙，粟粉 1 湯匙

1 tsp grated garlic, 1/2 tsp salt, ground black pepper, 1/2 tbsp sugar, 1 tsp oil, 1 tbsp water, 1 tbsp cornflour

調味 Seasoning

紅酒 1/4 杯，意大利黑醋 1/4 杯，喼汁 2 茶匙，
茄膏 1 湯匙，糖 1 湯匙，鹽少許

1/4 cup dry red wine, 1/4 cup balsamic vinegar, 2 tsps worcestershire sauce, 1 tbsp tomato paste, 1 tbsp sugar, salt

做法 Method

1. 豬扒洗淨，抹乾水分。用刀背拍鬆，下醃料醃約 30 分鐘。
2. 燒熱油，下豬扒用大火煎至兩面金黃色，轉用慢火煎熟，盛起備用。
3. 原鑊下油炒香蒜茸及洋蔥，再加入番茄，灑入少許鹽續炒，加入調味料拌勻煮滾，加入豬扒拌炒至汁液濃稠，上碟，以番茜裝飾即可。

1. Rinse pork chops and drain well. Pat lightly with the flat of knife. Mix with the marinade for 30 minutes.
2. Heat the oil. Fry pork chops over high heat until two sides are golden. Then reduce to low heat and fry until they are done. Set aside.
3. Stir fry grated garlic and onion in the same wok. Put in tomatoes and salt and stir fry. Add the seasoning and bring to the boil. Put in pork chops until the sauce is thickens. Arrange on the plate and garnish with parsley. Serve.

香茅洋蔥豬扒
Pork Chops with Lemongrass and Onion

不想香茅味太濃重，可酌減一半香茅份量。

In case you prefer a milder taste of lemongrass, cut its amount to half.

材料 Ingredients

豬扒 8 塊，香茅 2 棵，洋葱 1 個，葱段 1 棵

8 pieces pork chop, 2 stalks lemongrass, 1 onion,
1 sprig spring onion (sectioned)

醃料 Marinade

糖 1 茶匙，鹽半茶匙，老抽半茶匙，
胡椒粉少許

1 tsp sugar, 1/2 tsp salt, 1/2 tsp dark soy sauce,
ground white pepper

調味 Seasoning

魚露 2 湯匙，糖 1 茶匙，紹酒半茶匙

2 tbsps fish gravy, 1 tsp sugar, 1/2 tsp Shaoxing wine

做法 Method

1. 香茅斜切成片；洋葱切大粒。
2. 豬扒洗淨，瀝乾水分，用刀背拍鬆，用醃料
   醃 30 分鐘。
3. 豬扒沾上麵粉，下油鑊煎至熟。
4. 燒熱油，加入洋葱、香茅及葱炒香，放入豬
   扒及調味料，蓋上鍋蓋煮片刻，盛起即可。

1. Cut lemongrass into slices diagonally. Cut the
   onion into thick dices.
2. Wash and wipe dry pork chops. Pat slightly with
   the flat of the knife. Marinate pork chops for 30
   minutes.
3. Coat pork chops with cornflour. Fry pork chops
   until done.
4. Heat the oil. Stir fry onion, lemongrass and spring
   onion. Put in pork chops and seasoning. Cover the
   lid and cook for a while. Remove and serve.

雜 菜 焗 豬 扒 飯
Baked Pork Chop Rice
with Assorted Vegetables

若想豬扒飯更惹味,白飯可先跟雜菜同炒;想更健康
一點,則宜分開烹調。

For an extra dimension of flavour, you may fry the rice with
the assorted vegetables first. Yet, if you prefer a healthier diet,
you should cook the rice and assorted vegetables separately.

材料
Ingredients

豬扒 2 塊，雜菜半碗，西蘭花少許，荷蘭豆半碗，芝士 5 片，洋蔥半個，白飯 2 碗，鹽少許，雞蛋 1 個（拂匀），生粉 5 湯匙

2 pieces pork chop, 1/2 bowl mixed vegetable, broccoli, 1/2 bowl snow peas, 5 slices cheese, 1/2 onion, 2 bowls rice, salt, 1 egg (beaten), 5 tbsps caltrop starch

醃料
Marinade

生抽 1 1/2 茶匙，粟粉 1 茶匙，胡椒粉少許

1 1/2 tsps light soy sauce, 1 tsp cornflour, ground white pepper

做法
Method

1. 荷蘭豆、洋蔥切粒。燒熱油，下洋蔥炒香，加入雜菜、西蘭花及荷蘭豆，灑少許鹽炒匀，盛起，備用。

2. 豬扒洗淨，抹乾水分，用刀背拍鬆，下醃料醃約 1 小時。豬扒沾上蛋液，再拍上生粉，炸至金黃色，備用。

3. 預熱焗爐至 200℃，將白飯放入焗盤內，放上豬扒、雜菜料、芝士，放入焗爐焗 15 至 20 分鐘，即可食用。

1. Cut snow peas and onion into dice. Heat the oil. Stir fry onion until frangant. Add mixed vegetable, broccoli, snow peas and sprinkle salt, mix well. Set aside.

2. Rinse pork chops and drain well. Pat lightly with the flat of knife. Mix with the marinade for 1 hour. Mix with egg mixture and coat with caltrop starch. Heat the oil. Deep fry the pork chops until golden brown. Set aside.

3. Preheat an oven to 200°C. Put the rice into baking bowl. Add pork chops, assorted vegetable and cheese on top. Bake 15-20 minutes. Serve.

蜜 桃 芥 末 豬 扒
Apricot and Mustard Pork Chops

豬扒 8 塊，蜜桃果醬 1 杯，Dijon芥末醬 3 湯匙

8 pieces pork chop, 1 cup apricot jam,
3 tbsps Dijon mustard

鹽及黑胡椒碎各適量

salt and ground black pepper

1. 豬扒洗淨，瀝乾水分。用刀背拍鬆，與醃料拌勻醃 30 分鐘。
2. 豬扒煎至熟，盛起及備用。
3. 蜜桃果醬及芥末醬拌勻，用小火煮至果醬至溶（若醬料太濃稠，可酌量加入水分），放入豬扒輕拌，即可享用。

1. Wash and wipe dry pork chops. Pat lightly with the flat of the knife. Marinate pork chops for 30 minutes.
2. Fry pork chops until done. Remove and set aside.
3. Mix apricot jam and mustard. Cook the mixture over low heat until it is melted (add some water if the sauce is too sticky). Put in pork chops and mix until pork chops coat with apricot mustard sauce. Serve.

若找不到Dijon芥末醬，可用其他芥末醬代替。

If you can't get Dijon mustard, you may use other mustard instead.

如果是新鮮及厚身豬扒，宜醃一整夜；若是急凍及薄身的豬扒，醃數小時即可。

Marinate them the night before if you use fresh and thick pork chops. If you use frozen and thinly sliced pork chops, you just need to marinate them for a few hours.

海 鮮 醬 豬 扒
Pork Chops in Hoisin Sauce

材料 / Ingredients

豬扒 4 塊，蜂蜜 1 茶匙
4 pieces pork chop, 1 tsp honey

醃料 / Marinade

海鮮醬 3 湯匙，生抽 1 茶匙，粟粉 1 茶匙
3 tbsps Hoisin sauce, 1 tsp light soy sauce,
1 tsp cornflour

做法 / Method

1. 豬扒洗淨及抹乾，用醃料醃一整晚。
2. 燒熱平底鑊，用小火將豬扒煎至熟透。加入
   蜂蜜拌勻調味，即可食用。

1. Wash and wipe dry pork chops. Mix pork chops
   with marinade overnight.
2. Heat oil in pan. Fry pork chops on low heat until
   done. Season with honey. Serve.

日 式 咖 喱 豬 扒 飯
Pork Chops Rice in Japanese Curry

豬扒 2 塊，紅蘿蔔半個，洋葱半個，
日式咖喱磚 4 塊，水半杯，熱飯 1 至 2 碗
2 pieces pork chop, 1/2 carrot, 1/2 onion, 4 cubes
Japanese curry, 1/2 cup water, 1-2 bowls cooked rice

1. 洋葱切片；紅蘿蔔洗淨，切片。
2. 燒熱油，下洋葱及紅蘿蔔爆香，盛起備用。
3. 燒熱油，下豬扒煎熟至金黃色，下洋葱及紅
   蘿蔔，加入咖喱磚及水煮至汁液濃稠，盛起
   伴熱飯，趁熱進食。

1. Slice the onion. Rinse and slice the carrot.
2. Heat the oil. Put in onion and carrot until fragrant.
   Set aside.
3. Heat the oil. Fry the pork chops until done and
   golden. Put in cooked onion and carrot and stir
   fry. Add curry cubes and water and cook until the
   sauce thickens. Serve with the cooked rice.

酥 炸 豬 扒
Deep Fried Pork Chops

炸豬扒時，宜先用小火，油溫不能過
熱，否則容易炸至焦燶。

Always deep fry the pork chops over low
heat. They burn easily if the oil is too hot.

豬扒 8 塊，麵包糠 2 杯，雞蛋 2 個（拂勻），
麵粉少許

8 pieces pork chop, 2 cups breadcrumbs,
2 egg (beaten), flour

鹽 1 茶匙，糖 1 茶匙，蒜頭 2 粒（磨碎），
洋葱半個（切碎），白酒 3 茶匙，胡椒碎少許

1 tsp salt, 1 tsp sugar, 2 cloves garlic (crushed),
1/2 onion (finely chopped), 3 tsps white wine,
crushed black pepper

1. 豬扒洗淨，瀝乾水分，用醃料拌勻醃 1 小時。
2. 用適量麵粉沾在豬扒上，再沾上蛋液，最後加
   上麵包糠。
3. 燒熱油，放入豬扒炸至熟透，可蘸沙律醬食
   用。

1. Wash and wipe dry pork chops. Marinate for 1 hour.
2. Coat pork chops with some flour. Dip pork chops in
   egg wash and then coat in breadcrumbs.
3. Deep fry pork chops in hot oil until done. Serve hot
   with salad dressing.

蜜 汁 煎 火 腿 豬 扒
Fried Pork Chops and Jinhua Ham with Honey

無骨豬扒 8 塊，金華火腿 3 兩，菊花蜜 6 湯匙，
生粉 4 茶匙

8 pieces boneless pork chop, 113 g Jinhua ham,
6 tbsps chrysanthemum honey, 4 tsps caltrop starch

生抽 2 茶匙，老抽 1 茶匙
2 tsps light soy sauce, 1 tsp dark soy sauce

1. 金華火腿去皮、切片，澆上菊花蜜 2 湯匙，隔水
   蒸約 15 分鐘，備用。
2. 豬扒洗淨、切件，用刀背略剁鬆，撒上生粉醃
   15 分鐘，令豬扒更軟滑。
3. 燒熱鑊，下油用慢火將豬扒煎至兩邊熟透，加入
   火腿片略煎。
4. 菊花蜜 4 湯匙及調味料調勻，澆在肉面煎封拌勻
   即可。

1. Skin and slice the ham. Pour over 2 tbsps of honey and
   steam for 15 minutes. Set aside.
2. Rinse pork chops and cut into pieces. Pat lightly with
   the flat of the knife. Sprinkle with caltrop starch and
   leave for 15 minutes to make them soft.
3. Heat the oil. Fry the pork chops until done and golden.
   Put in the ham and fried lightly.
4. Mix 4 tbsps of honey and seasoning well. Pour over the
   pork and ham and cook for a while. Place on the plate
   and serve.

宜選外型飽滿及體大的番茄，以
免味道過酸。

Pick tomatoes that are plump and large.
Otherwise, they might be too sour.

鮮 茄 蛋 豬 扒
Pork Chops with Tomato and Fried Egg

材料 Ingredients

豬扒 3 塊，大番茄 2 個，雞蛋 3 至 4 個

3 pieces pork chop, 2 big tomatoes, 3-4 eggs

醃料 Marinade

生抽 1 湯匙，糖半茶匙，胡椒粉少許，
粟粉 1 茶匙

1 tbsp light soy sauce, 1/2 tsp sugar, ground white pepper, 1 tsp cornflour

調味 Seasoning

茄汁半碗，糖半茶匙，生抽半茶匙

1/2 bowl ketchup, 1/2 tsp sugar, 1/2 tsp light soy sauce

做法 Method

1. 豬扒洗淨及瀝乾，用刀背拍鬆，用醃料醃 1 小時。
2. 番茄洗淨及切角。
3. 雞蛋拂打均勻，炒成蛋塊，備用。
4. 豬扒煎至熟，切成 3 塊，備用。
5. 燒熱油 1 茶匙，加入番茄及水炒片刻至熟透，下豬扒、蛋塊及調味料炒勻，即可食用。

1. Wash and drain the pork chops. Pat lightly with the flat of knife. Mix with marinade for at least 1 hour.
2. Wash tomatoes and cut into quarters.
3. Whisk egg and stir fry. Remove and set aside.
4. Fry the pork chops until done. Cut each into 3 pieces. Set aside.
5. Heat 1 tsp of oil, put in the tomatoes and water and cook until done. Add the pork chops, eggs and seasoning. Stir fry for a while. Serve.

蘑菇香草豬扒

Pork Chops with Button Mushrooms and Rosemary

若找不到新鮮迷迭香，可用乾品代替。

If you can't get any fresh rosemary, use dried one instead.

材料 Ingredients

無骨豬扒 4 塊，洋蔥半個（切粒），
蘑菇 1 罐，迷迭香適量，蒜頭 3 粒（剁碎）

4 pieces boneless pork chop, 1/2 onion (diced),
1 can button mushroom, rosemary,
3 cloves garlic (chopped)

醃料 Marinade

生抽 1 茶匙，糖 1 1/2 茶匙，粟粉少許

1 tsp light soy sauce, 1 1/2 tsps sugar, cornflour

調味 Seasoning

麻油1 1/2 茶匙，胡椒粉少許

1 1/2 tsps sesame oil, ground white pepper

做法 Method

1. 豬扒切成小片，用醃料醃 1 小時。
2. 蘑菇及洋蔥炒香，備用。
3. 燒熱油，下蒜茸、豬扒及調味料炒香，下迷迭
   香略煮，最後放入蘑菇及洋蔥，即可食用。

1. Cut pork chops into pieces and marinate for 1 hour.
2. Stir fry mushrooms and onion. Set aside.
3. Heat oil. Stir fry grated garlic, pork chops and
   seasoning. Add rosemary and cook for a while. Put
   in mushrooms and onion lastly. Serve.

檸汁豬扒
Pork Chops in Lemon Sauce

材料 Ingredients

豬扒（梅頭豬肉）4 塊，檸檬 3 片，雞蛋 1 個，麵粉約半杯

4 pieces pork chop, 3 slices lemon, 1 egg, 1/2 cup flour

醃料 Marinade

鹽、糖各半茶匙，生抽 1 茶匙，生粉 1 茶匙，水 1 湯匙，胡椒粉少許

1/2 tsp salt, 1/2 tsp sugar, 1 tsp light soy sauce, 1 tsp caltrop starch, 1 tbsp water, ground white pepper

獻汁 Sauce

糖 1 湯匙，生粉 1 茶匙，檸檬汁 3 湯匙，水半杯

1 tbsp sugar, 1 tsp caltrop starch, 3 tbsps lemon juice, 1/2 cup water

做法 Method

1. 豬扒洗淨，用刀背拍鬆，以醃料拌勻略醃。
2. 雞蛋拂勻，豬扒沾上蛋液後，再均勻地裹上麵粉，以中火炸熟，瀝乾油分，排放碟上。
3. 煮滾獻汁，放入檸檬片煮一會，澆在豬扒上即成。

1. Rinse pork chops. Pat lightly with the flat of the knife. Mix with marinade for a while.
2. Whisk the egg. Coat pork chops lightly with the egg wash. Brush the flour evenly. Heat the oil. Deep fry the pork chops over medium heat until done. Drain the oil. Remove to the plate.
3. Bring the sauce to the boil. Add the lemon slices and cook for a while. Pour the sauce over the pork chops and serve.

獻汁的份量要小心控制，因會影響菜式的酸甜度，用少量炸豬扒油與獻汁拌勻，味道更佳。

Make sure you control the volume of the sauce carefully. Too much or too little sauce would affect the balance between sourness and sweetness directly. For the best result, cook the sauce with a little oil used to deep frying the pork chops.

南 瓜 咖 喱 雞 翼
Curry Chicken with Pumpkin

雞中翼 1 斤，南瓜 1 斤，罐裝紅腰豆 1
罐（約300克），咖喱醬 4 湯匙

600 g chicken wings, 600 g pumpkin, 300 g
canned red kidney beans, 4 tbsps curry paste

生抽 3 湯匙，糖 2 湯匙

3 tbsps light soy sauce, 2 tbsps sugar

1. 南瓜去皮、去籽，切粒。
2. 雞翼洗淨，抹乾水分，下醃料拌勻醃半小
   時。
3. 燒熱油，下雞翼略煎，再放入南瓜粒略
   炒，加入紅腰豆、咖喱醬及水 1 杯，加蓋
   煮半小時，待雞翼入味及汁液濃稠，即可
   享用。

1. Peel and seed pumpkin. Cut into cubes.
2. Rinse chicken wings. Wipe dry. Mix with the
   marinade for half an hour.
3. Heat the oil. Fry chicken wings slightly. Put in
   pumpkin and stir fry. Add red kidney beans,
   curry paste and 1 cup of water. Cover with the
   lid and cook for 30 minutes until the chicken
   wings done and the sauce thickens. Serve.

水 牛 雞 翼
Buffalo Chicken Wings

傳統的水牛雞翼可配藍芝士醬汁
和西芹條一併吃，滋味無窮。

Buffalo chicken wings are traditionally
served with blue cheese dressing and
celery sticks.

材料 Ingredients

雞全翼 1 1/2 斤至 2 斤 4 兩，
植物油（或牛油）2 湯匙

900 g-13.5 kg whole chicken wings,
2 tbsps vegetable oil or butter

 醃料 Marinade

黑胡椒碎適量，鹽適量
ground black pepper, salt

 調味 Seasoning

辣椒醬或辣椒汁 2 1/2 湯匙，白酒醋 1 湯匙
2 1/2 tbsps chilli sauce or tabasco sauce,
1 tbsp white wine vinegar

 做法 Method

1. 將雞中翼與雞小槌切開（雞翼尖切掉），塗上黑
   胡椒碎，若有需要可灑上鹽。
2. 燒熱油，下半份雞翼，不時翻動，炸至金黃及香
   脆，待雞翼熟透，盛起，放在廚房紙上瀝乾，再
   加入餘下的雞翼炸至熟。
3. 燒熱植物油，加入辣椒醬及白酒醋拌勻，立即關
   火。
4. 將雞翼排放在溫熱的碟上，澆上步驟（3）的醬
   汁即可。

1. Chop up the mid-joint chicken wings and chicken
   drums (cut off the chicken tips). Grind on black
   pepper and sprinkle with salt if desired.
2. Heat the oil. Add half portion of chicken wings and
   cook until golden brown and crisp (stirring while deep
   frying). Drain the oil on the kitchen paper. Cook the
   remaining chicken wings.
3. Heat the vegetable oil, add the chilli sauce and white
   wine vinegar. Stir well and remove from the heat.
4. Place the chicken on a warm serving platter, pour the
   sauce from steps (3) on chickens wings and serve.

冬 菇 紅 棗 燜 雞 翼

Stewed Chicken Wings with Mushrooms and Red Dates

材料 Ingredients

雞翼 10 隻，冬菇 8 朵（浸軟），
紅棗 8 粒（去核），薑 3 片

10 chicken wings, 8 dried black mushrooms (soaked until soft), 8 red dates (stoned), 3 slices ginger

醃料 Marinade

生抽 1 茶匙，糖、胡椒粉、粟粉各少許

1 tsp light soy sauce, sugar, ground white pepper, cornflour

調味 Seasoning

蠔油 2 湯匙，生抽 1 茶匙，老抽 1 湯匙，
冰糖碎 1 湯匙

2 tbsps oyster sauce, 1 tsp light soy sauce,
1 tbsp dark soy sauce, 1 tbsp rock sugar bits

做法 Method

1. 雞翼用醃料醃約 1 小時。
2. 燒熱油 2 湯匙，爆香薑片和冬菇，下雞翼和紅棗，兜炒一會後，下調味料和水 3/4 杯，煮滾後改調小火，煮至汁液濃稠即可享用。

1. Marinate chicken wings for 1 hour.
2. Heat 2 tbsps of oil. Stir fry sliced ginger and mushrooms until fragrant. Add chicken wings and red dates. Stir fry for a while. Add seasoning and 3/4 cup of water. Bring to the boil and turn to low heat. Cook until sauce thickens. Ready to serve.

泰 式 煎 雞 翼
Fried Chicken Wings in Thai Sauce

建議將雞翼放入膠袋內與醃料拌勻，
令雞翼醃得均勻，而且方便儲存。

You may put the chicken wings into a zipper
bag. Pour in the marinade and mix well.
All chicken wings will be soaked in the
marinade evenly. They can also be stored in
the fridge more conveniently.

雞中翼 12 隻，番茜 1 棵（切碎），
葱 1 條（切碎）

12 chicken wings, 1 sprig coriander (finely chopped),
1 sprig spring onion (finely chopped)

鹽 1 湯匙，黑胡椒粉少許

1 tbsp salt, ground white black pepper

青檸 2 個（榨汁），魚露 2 湯匙，糖 4 1/2 湯
匙，蕎頭 8 粒（切碎），蒜頭 8 粒（切粒），
紅辣椒 3 隻（切圈）

2 lime (squeezed), 2 tbsps fish gravy, 4 1/2 tbsps sugar,
8 pickled bulbous onion (finely chopped), 8 cloves
garlic (diced), 3 red chilies (cut into rings)

1. 雞翼洗淨，抹乾水分，用醃料醃 30 分鐘。
2. 雞翼拍上適量粟粉，下油鑊煎至熟，備用。
3. 煮滾調味料，加入雞翼，灑入番茜及青葱拌
   勻，即可食用。

1. Rinse the chicken wings and wipe dry. Marinate
   and leave it for 30 minutes.
2. Sprinkle some cornflour on the chicken wings. Fry
   chicken wings until done. Set aside.
3. Bring the seasoning ingredients to the boil. Put the
   chicken wings back and sprinkle coriander and
   spring onion. Stir well and serve.

洋 葱 雞 翼

Stir Fried Chicken Wings with Onion

材料 Ingredients

雞中翼 8 隻，洋蔥半個

8 chicken wings, 1/2 onion

醃料 Marinade

生抽 1 湯匙，老抽 1 茶匙，蠔油 1 茶匙，
冰糖 1 塊（切碎），胡椒粉少許

1 tbsp light soy sauce, 1 tsp dark soy sauce,
1 tsp oyster sauce, 1 piece rock sugar (finely chopped),
ground white pepper

調味 Seasoning

喼汁 2 湯匙

2 tbsps worcestershire sauce

做法 Method

1. 雞翼洗淨，與醃料拌勻，加少許水待半小時
   （醃汁留用）。
2. 洋蔥洗淨，切粗絲備用。
3. 燒熱油，用中火略煎洋蔥及雞翼，雞翼煮至金
   黃色，加入步驟（1）的醃汁料同煮，最後
   加上喼汁，煮熟後即可上碟。

1. Rinse chicken wings. Stir with the marinade. Pour a
   little water and leave them for 30 minutes (keep the
   marinade for later use).
2. Rinse and thickly shred the onion. Set aside.
3. Heat the oil. Fry chicken wings and onion together
   over medium heat. Cook the chicken wings until
   golden brown. Pour in the marinade sauce from
   step (1) and cook a while. Add the seasoning at last.
   Bring to the boil. Serve.

酥 炸 小 雞 槌
Crispy Chicken Mallets

材料 Ingredients

小雞槌 10 隻，雞蛋 1 個（拂勻），麵粉 1 杯
10 chicken mallets, 1 egg (beaten), 1 cup flour

醃料 Marinade

鹽半茶匙，紹酒半茶匙，生抽 1 湯匙，
老抽 1 茶匙，糖 1 茶匙
1/2 tsp salt, 1/2 tsp Shaoxing wine, 1 tbsp light soy
sauce, 1 tsp dark soy sauce, 1 tsp sugar

做法 Method

1. 小雞槌洗淨，抹乾水分，用醃料醃 2 小時。
2. 小雞槌先沾上蛋液，再拍上麵粉，放入熱油
   內炸熟，盛起，即可趁熱享用。

1. Wash and wipe dry chicken mallets. Marinate for 2
   hours.
2. Dip the chicken mallets first in egg wash and then
   coat in flour. Deep fry in hot oil until done. Serve
   hot.

由於雞槌肉厚，想縮短醃味時間，用叉子在厚肉部位刺入，易於入味。炸時先用小火，將炸熟時轉大火，令雞槌外層更金黃香脆。

The chicken mallets are thick and fleshy. To shorten the marinating time, you may pierce through the thick meat with a fork before adding marinade. Besides, deep-fry them over low heat. Then turn it up to high heat when they are almost done.

豉油王雞翼
Chicken Wings in Soy Sauce

此食譜必須用冰糖烹調，令雞翼容易熟透，而且顏色較美觀，但烹煮時間不宜過長。

Rock sugar makes the chicken wings cook more quickly and adds a caramel colour to the wings after cooked. However, do not cook them for too long.

材料 Ingredients

雞中翼 10 隻

10 mid-joint chicken wings

調味 Seasoning

葱粒、薑粒各少許，老抽 3 湯匙，生抽 3 湯匙，水 3 至 6 湯匙（可自行調校濃淡度），冰糖 1 塊，白酒少許

ginger and spring onion, 3 tbsps dark soy sauce, 3 tbsps light soy sauce, 3-6 tbsps water (depends on your taste), 1 piece rock sugar, white wine

做法 Method

1. 燒滾水，放入雞翼飛水及煮至七成熟，放於水喉下啤水約 1 分鐘，去掉油分。

2. 煮滾調味料，放入雞翼，待汁料再滾起，立即關火，待雞翼浸約半小時，即可食用。

1. Bring the suitable amount of water to the boil. Scald chicken wings until half done. Rinse under the tap water to remove the oil.

2. Bring the seasoning ingredients to the boil. Put in chicken wings. When the sauce boils again, turn off the heat. Soak the chicken wings in sauce for half an hour. Ready to serve.

蝦醬酥炸雞翼
Deep Fried Chicken Wings with Shrimp Paste

材料 Ingredients

雞中翼約 10 隻，生粉適量

10 mid-joint chicken wings, caltrop starch

醃料 Marinade

蝦醬 1 湯匙，蒜茸半湯匙，糖半茶匙，
生抽半湯匙，薑汁半茶匙，紹酒半茶匙，
胡椒粉、麻油、水各少許

1 tbsp shrimp paste, 1/2 tbsp grated garlic, 1/2 tsp sugar, 1/2 tbsp light soy sauce, 1/2 tsp ginger juice, 1/2 tsp Shaoxing wine, ground white pepper, sesame oil, water

做法 Method

1. 雞中翼洗淨，吸乾水分，下醃料醃 30 分鐘
   至入味，雞翼抹上生粉。
2. 燒熱油，以中火炸雞翼至金黃香脆，趁熱
   享用。

1. Wash and wipe dry chicken wings. Marinate for 30 minutes and coat with caltrop starch.
2. Heat the oil. Deep fry chicken wings until golden brown. Serve hot.

炸蛋木耳燜雞翼

Stewed Chicken Wings with Deep Fried Eggs and Wood Ear Fungus

焗雞蛋時，建議水與雞蛋一同開火煮，並加少許鹽，雞蛋殼不容易爆裂。

To hard boil the eggs, put the eggs into a pot with cold water and add a pinch of salt before heating it up on the stove. The eggs are less likely to crack this way.

材料 Ingredients

雞翼 12 隻，雞蛋 6 個，木耳 3 朵，
玫瑰露酒及清水各適量

12 chicken wings, 6 eggs, 3 wood ear fungus,
rose wine, water

配料 Condiments

乾葱 4 粒，蒜肉 4 粒，薑 3 片，八角 2 粒，
桂皮 1 片

4 cloves shallot, 4 cloves garlic, 3 slices ginger,
2 star aniseeds, 1 piece cassia bark

調味 Seasoning

生抽半杯，鹽 1 茶匙，片糖半塊，麻油 1 茶匙

1/2 cup light soy sauce, 1 tsp salt,
1/2 slab brown sugar, 1 tsp sesame oil

做法 Method

1. 雞蛋焓熟、去殼，抹乾水分。放入滾油內炸至金黃色，備用。
2. 木耳浸透，洗淨，去硬蒂、切片，瀝乾水分。
3. 燒熱油 1 湯匙，爆香配料，下雞翼，注入過面清水，以中慢火燜 10 分鐘，加入木耳、調味料、玫瑰露酒再燜 10 分鐘，最後加入雞蛋再燜片刻即可。

1. Poach eggs until hard. Shell and wipe dry. Deep fry the eggs until golden brown. Set aside.
2. Soak wood ear until soft. Wash and remove the hard stem. Cut into pieces and drain.
3. Heat 1 tbsp of oil. Stir fry the condiments until fragrant. Put in chicken wings and water (must cover the chicken wings). Simmer for 10 minutes over medium-low heat. Add wood ear, seasoning and rose wine. Simmer for 10 more minutes. Put in eggs and simmer for a while. Serve.

滋味雞翼
Chicken Wings
in Worcestershire Sauce

雞翼 1 1/2 斤，乾蔥 2 粒，水適量

900 g chicken wings, 2 cloves shallot, water

喼汁 6 湯匙，糖 3 湯匙，老抽 1 湯匙，
鹽少許

6 tbsps worcestershire sauce, 3 tbsps sugar,
1 tbsp dark soy sauce, salt

1. 雞翼洗淨，抹乾水分，下醃料拌勻醃約半小時（醃料汁留用）。
2. 燒熱油，爆香乾蔥，下雞翼炒一會，下醃料汁及水（水要蓋過雞翼），煮約 15 分鐘即成。

1. Wash and wipe dry chicken wings. Marinate for 30 minutes (leave the marinade for later use).
2. Heat the oil in wok. Stir fry shallots until fragrant. Put in chicken wings and stir fry. Add marinade and water (water level should be covered the ingredients). Cook for 15 minutes. Serve.

先爆香乾葱，能增加這道菜式
的惹味程度。

Stir-fry the shallot until fragrant first
before putting in other ingredients.
This step adds a pleasing aroma to
the dish.

豆豉雞翼煲

Simmered Chicken Wings with Fermented Black Beans in Claypot

燜雞翼時，爐火不要太猛，否則雞肉會硬，影響口感。

Do not stew the chicken wings over high heat. Otherwise, the meat will be dried out.

雞中翼 10 至 12 兩，乾蔥 10 粒，薑 2 片，
原粒豆豉 3/4 湯匙，蔥粒 1 湯匙

375 g-450 g mid-joint chicken wings,
10 cloves shallot, 2 slices ginger, 3/4 tbsp fermented
black beans (in whole), 1 tbsp diced spring onion

醃料 Marinade

薑汁半茶匙，紹酒半茶匙，生抽半湯匙，
粟粉、胡椒粉各適量

1/2 tsp ginger juice, 1/2 tsp Shaoxing wine,
1/2 tbsp light soy sauce, cornflour,
ground white pepper

調味 Seasoning

水半杯，鹽 1/8 茶匙，糖 1 茶匙，
生抽 3/4 湯匙，麻油、胡椒粉各適量

1/2 cup water, 1/8 tsp salt, 1 tsp sugar, 3/4 tbsp light
soy sauce, sesame oil, ground white pepper

做法 Method

1. 雞翼洗淨，斬件，加入醃料醃勻待半小時，泡
   油備用。
2. 乾蔥去衣，洗淨，泡油備用。
3. 燒熱砂鍋，加入油約 1 1/2 湯匙，爆香薑片及
   豆豉粒，下雞翼及乾蔥炒勻，灒入紹酒，注入
   調味料煮滾，改用中小火稍煮片刻至材料熟透
   及汁料濃稠，最後灑上蔥粒，原鍋享用。

1. Rinse and chop chicken wings into pieces. Stir well
   with the marinade for half an hour. Scald in hot oil
   for a while.
2. Skin and rinse the shallots. Scald in hot oil for a
   while.
3. Heat the clay pot. Add 1 1/2 tbsps of oil. Stir fry
   ginger slices and fermented black beans until
   fragrant. Put in chicken and shallots. Sprinkle
   wine and seasoning. Bring to the boil and turn to
   medium-low heat. Cook until chicken wings done
   and the sauce thickens. Add diced spring onion.
   Serve with the clay pot.

用牙籤緊緊穿起雞翼，令釀入材料的
雞翼不容易散開。

To prevent the stuffed wings from bursting,
seal the seam securely with toothpicks or
bamboo skewers.

鮮 菇 甜 椒 釀 雞 翼
Stuffed Chicken Wings with
Mushrooms and Bell Peppers

材料 Ingredients

雞中翼 12 隻，鮮冬菇 4 朵，紅甜椒 1 個，
蘆筍 6 枝，蒜茸 1 茶匙

12 mid-joint chicken wings, 4 fresh black mushrooms,
1 red bell pepper, 6 stalks asparagus, 1 tsp minced garlic

醃料 Marinade

生抽 2 湯匙，胡椒粉少許，粟粉 1 茶匙

2 tbsps light soy sauce, ground white pepper, 1 tsp cornflour

調味 Seasoning

清雞湯半杯，清水
半杯，老抽 2 茶
匙，糖 1 茶匙
1/2 cup chicken
stock, 1/2 cup water,
2 tsps dark soy sauce,
1 tsp sugar

獻汁 Cornflour Solution

水 1 湯匙，
粟粉 1 茶匙 *拌勻
1 tbsp water, 1 tsp
cornflour *mixed

做法 Method

1. 雞中翼洗淨，去骨，翻出雞肉，下醃料醃約半小時。
2. 鮮冬菇每朵切成 3 件；紅甜椒切成 12 條；蘆筍切半。
3. 將鮮冬菇、紅甜椒、蘆筍各 1 條釀入已塗粟粉的雞翼內。
4. 燒熱油 1 湯匙，下雞翼略煎兩面，注入調味料，用慢火煮 10 分鐘，埋獻上碟即可。

1. Rinse and bone mid-joint chicken wings. Turn the flesh inside out. Marinate for half an hour.
2. Cut each mushroom into 3 pieces. Cut bell pepper into 12 strips. Halve the asparagus.
3. Stuff each chicken wing with 1 piece of fresh black mushroom and 1 strip of red bell pepper and asparagus.
4. Heat 1 tbsp of oil in a wok. Fry both sides of the chicken wings briefly. Pour in seasoning. Cook over low heat for 10 minutes. Stir in cornflour solution. Stir well and serve.

白雲雞翼尖
Chicken Wing Tips in White Spice Sauce

煮熱雞翼尖，再用冷水沖洗，令外皮爽彈脆口。此道菜冷熱吃皆宜。

After you heat up the chicken wing tips, rinse them in cold or iced water. That would make their skin more crunchy and less sticky. This dish can be served hot or cold.

雞翼尖 1 1/2 斤，薑 2 片，蔥 1 條（切段），
番茜 1 棵，紅辣椒 1 隻（切片），麻油少許

900 g chicken wing tips, 2 slices ginger, 1 sprig spring
onion (sectioned), 1 sprig parsley, 1 red chilli (sliced),
sesame oil

汁料 Sauce

花椒粒 1 茶匙，八角 2 粒，陳皮 1/4 個，鹽
1 1/2 茶匙，糖半茶匙，紹酒 1 湯匙，水 5 杯

1 tsp Sichuan peppercorns, 2 star aniseeds, 1/4 dried
tangerine peel, 1 1/2 tsps salt, 1/2 tsp sugar, 1 tbsp
Shaoxing wine, 5 cups water

做法 Method

1. 將花椒粒、八角及陳皮放入布袋內，放入滾水
   內煲半小時。把布袋拿走，加入鹽、糖及紹
   酒，煮滾備用。
2. 雞翼尖洗淨及瀝乾水分，抹上鹽，用冷水沖洗
   乾淨。
3. 燒滾一鍋水，下薑及蔥，放入雞翼尖煮 5 分
   鐘，取出，用冷水沖淨。
4. 雞翼尖放入步驟（1）的汁料內，燜煮 10 分
   鐘（勿加上蓋），取出，待涼，放入雪櫃。吃
   時灑上麻油、番茜及紅辣椒即可。

1. Put Sichuan peppercorns, star aniseed and dried
   tangerine peel in a muslin bag and boil in water for
   half an hour. Discard the bag. Add salt, sugar and
   wine. Bring the liquid to the boil and set aside.

2. Wash and drain chicken wing tips. Rub with salt.
   Then rinse under the tap water.

3. Blanch chicken wing tips in boiling water with
   ginger and spring onion for 5 minutes. Remove and
   rinse them under the tap water.

4. Put chicken wing tips into white spice sauce from
   step (1). Simmer for 10 minutes (without cover with
   the lid). Allow to cool, then put into a refrigerator.
   Sprinkle with sesame oil, parsley and red chilli.
   Serve.

若雞翼肉較厚，建議醃一晚，容易入味。

If you use large wings with thick flesh, marinate them the night before for them to pick up more of the flavour.

甜豉油雞翼
Chicken Wings in Sweet Soy Sauce

材料 Ingredients

雞中翼約 10 隻

10 mid-joint chicken wings

醃料 Marinade

冰糖 2 至 3 湯匙（切碎），生抽 4 至 5 湯匙

2-3 tbsps rock sugar (finely chopped),
4-5 tbsps light soy sauce

做法 Method

1. 雞翼用醃料醃一晚。
2. 將雞翼放於墊上錫紙的焗盤內。
3. 預熱焗爐至 200℃，放入焗盤焗約 20 至 30 分鐘，再以 150℃至175℃焗至熟透，即可享用。

1. Marinate chicken wings for about overnight.
2. Spread aluminum foil on the baking tray. Put the chicken wings on the baking tray.
3. Preheat the oven at 200°C. Bake the chicken wings for 20-30 minutes. Then turn to 150-175°C and bake until the chicken wings done. Serve.

想雞翼的味道濃烈，可醃一整晚。

If you like stronger flavour, you may marinate them the night before.

甜 辣 雞 中 翼
Sweet and Spicy Chicken Wings

材料 **Ingredients**

雞中翼 12 隻，香橙果醬半杯，
辣椒粉 1 至 2 茶匙

12 mid-joint chicken wings,
1/2 cup orange marmalade,
1-2 tsps chilli powder

做法 **Method**

1. 香橙果醬、辣椒粉及雞翼同放入三文
   治袋內，搖晃至均勻，醃片刻。

2. 預熱焗爐至 180℃，將雞翼放在已墊
   錫紙的焗盤上，焗 25 分鐘或至熟，
   即可食用。

1. Mix marmalade and chilli powder in a
   plastic sandwich bag. Add chicken wings
   and shake until evenly coated for a while.

2. Preheat oven to 180°C. Place chicken
   wings on a baking tray lined with
   aluminium foil. Bake for 25 minutes or
   until done. Serve.

照燒牛柳粒
Stir Fried Beef Dices in Teriyaki Sauce

材料 Ingredients

牛柳 10 兩 (切粒)

400 g beef fillet (diced)

醃料 Marinade

照燒汁 2 茶匙，日式豉油 2 茶匙，糖適量

2 tsps Teriyake sauce, 2 tsps Japanese soy sauce, sugar

做法 Method

1. 牛肉粒用糖、日式豉油醃片刻，再拌入照燒汁（醃料留用）。
2. 燒熱油，下牛肉粒炒至半熟。拌入醃料汁，煮一會後盛起，即可食用。

1. Marinate diced beef with sugar and Japanese soy sauce, then mix with Teriyake sauce (keep the marinade for later use).
2. Heat some oil. Stir fry the diced beef until half done. Mix with the marinade sauce. Cook for a while. Remove and serve.

泰 式 牛 扒 沙 律
Beef Fillet Salad in Thai Style

不嗜牛者，可改用已炸脆的煙肉代替。

Those who don't eat beef may use deep-fried
smoked bacon instead of beef.

牛扒 2 塊，沙律菜 1 包，洋葱半個（切圈），粟米半碗（瀝乾水分），泰式甜辣醬適量，油 2 湯匙

2 pieces beef fillet, 1 pack salad vegetable, 1/2 onion (cut into ring), 1/2 bowl sweet corn kernels, Thai sweet chilli sauce, 2 tbsps oil

生抽、糖、粟粉各 1 茶匙

1 tsp light soy sauce, 1 tsp sugar, 1 tsp cornflour

1. 用醃料將牛扒醃 30 分鐘；沙律菜洗淨，瀝乾水分。
2. 燒熱油，下牛扒煎香，盛起，備用。
3. 將所有材料混合，以泰式甜辣醬伴食。

1. Mix beef fillet with the marinade and leave for 30 minutes. Rinse and drain salad vegetables.
2. Heat the oil. Fry beef fillet until fragrant. Set aside.
3. Mix all ingredients, then serve with Thai sweet chilli sauce.

家常炒粒粒
Stir Fried Diced Steak
with Vegetables

材料 Ingredients

西冷牛扒 1 塊（切粒），蒜頭 3 粒，菜脯 10 條（切粒），紅蘿蔔半條（切粒），蘆筍 6 條（切粒），沙葛半個（切粒）

1 beef steak (diced), 3 cloves garlic, 10 strips preserved radish (diced), 1/2 carrot (diced), 6 asparagus (diced), 1/2 yam bean (diced)

醃料 Marinade

糖 2 茶匙，老抽 1 茶匙，玫瑰露酒半茶匙，胡椒粉半茶匙，麻油 1 茶匙

2 tsps sugar, 1 tsp dark soy sauce, 1/2 tsp rose wine, 1/2 tsp ground white pepper, 1 tsp sesame oil

調味 Seasoning

玫瑰露酒、糖、鹽各半茶匙

1/2 tsp rose wine, 1/2 tsp sugar, 1/2 tsp salt

做法 Method

1. 燒熱鑊，下油爆香蒜頭 1 粒，加入紅蘿蔔炒一會，下蘆筍炒片刻，下少許水，蓋上鍋蓋煮一會，下菜脯和沙葛再兜炒一會，盛起。

2. 熱鑊下油，爆香蒜頭 2 粒，下牛柳粒煎至熟透，加入上述的雜菜料炒勻，下糖及鹽調味，炒勻，潷酒，炒勻即可上碟。

1. Heat wok and add oil. Stir fry 1 clove of garlic until fragrant. Put in carrot. Stir fry briefly. Add diced asparagus. Stir fry for a while. Add some water and cover the lid. Simmer for a while. Add preserved radish and yam bean. Stir fry for while and remove.

2. Heat wok and add oil. Stir fry 2 cloves of garlic. Put in diced steak. Fry the diced steak until done. Pour in diced vegetables from step (1). Stir well. Season with sugar and salt. Stir well. Sprinkle wine. Stir well and serve.

黑 椒 牛 柳
Fried Steak with Black Pepper

材料 Ingredients

牛扒 6 塊，牛油 1 茶匙
6 pieces steaks, 1 tsp butter

 醃料 Marinade

黑椒碎 1 茶匙，蒜鹽 1 茶匙，植物油 2 湯匙
1 tsp grated black pepper, 1 tsp garlic-salt,
2 tbsps vegetable oil

 醬汁 Sauce

黑椒碎半茶匙，乾葱 1 茶匙（切碎），
粟粉半茶匙，水 3 湯匙
1/2 tsp grated black pepper, 1 tsp chopped shallot,
1/2 tsp cornflour, 3 tbsps water

 做法 Method

1. 牛扒洗淨，瀝乾水分，用醃料拌勻醃 30 分鐘。
2. 燒熱平底鍋，加入牛油，下牛扒煎至半熟，盛
   起備用。
3. 預備醬汁：用小火把乾葱炒香，加入其他材料
   煮至滾。
4. 加入牛扒拌勻，即可食用。

1. Wash and wipe dry steaks. Marinate for 30 minutes.
2. Heat pan. Add butter. Put in steak and fry until
   medium-well done. Remove and set aside.
3. For sauce, stir fry chopped shallot over low heat until
   fragrant. Add other sauce ingredients. Bring to the
   boil.
4. Add steak and stir well. Serve.

黑椒牛柳絲蝴蝶粉

Farfalle with Sirloin Shreds in Black Pepper Sauce

煮滾蝴蝶粉後，不要再煮太
久，否則質感過軟。

Don't overcook the farfalle. Or
else it may be too soggy instead
of al dente.

材料 Ingredients

牛扒 1 斤（切絲），蝴蝶粉半包

600 g sirloin (shredded), 1/2 pack farfalle

醃料 Marinade

老抽 1 湯匙，油 1 茶匙，粟粉 1 茶匙，
麻油半茶匙

1 tbsp dark soy sauce, 1 tsp oil, 1 tsp cornflour,
1/2 tsp sesame oil

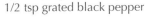

調味 Seasoning

蠔油半碗，水 1 碗，黑胡椒碎半茶匙

1/2 bowl oyster sauce, 1 bowl water,
1/2 tsp grated black pepper

做法 Method

1. 牛扒用醃料醃片刻。

2. 燒滾一鍋熱水，下鹽、油各 1 茶匙，放入蝴蝶粉，煮 8 至 10 分鐘，關火置 10 分鐘。將蝴蝶粉倒出，隔去多餘水分，備用。

3. 用平底鍋燒熱油 1 湯匙，加入牛柳絲炒勻，下調味料，炒勻後加入蝴蝶粉煮至熟透，待所有醬汁拌勻，即可食用。

1. Marinate with sirloin for a while.

2. Bring a pot of water to the boil. Put 1 tsp of salt and 1 tsp of oil in pot. Put the farfalle into the boiling water. Cook for 8-10 minutes. Turn off the heat and leave for 10 minutes. Pour into the strainer. Set aside.

3. Put 1 tbsp of oil into pan. Stir fry the sirloin shreds and mix with the seasoning sauce. Add the farfalle to cook until done. Mix the sauce with all ingredient evenly. Serve.

彩椒炒牛扒
Stir Fried Sirloin and Bell Peppers

材料 Ingredients

牛扒 8 兩，紅甜椒半個，青甜椒半個，
黃甜椒半個，蒜茸 1 茶匙

300 sirloin, 1/2 red bell pepper, 1/2 green bell pepper,
1/2 yellow bell pepper, 1 tsp grated garlic

醃料 Marinade

生抽 1 茶匙，麻油 1/4 茶匙，糖半茶匙，
粟粉半湯匙，胡椒粉少許

1 tsp light soy sauce, 1/4 tsp sesame oil, 1/2 tsp sugar,
1/2 tbsp cornflour, ground white pepper

調味 Seasoning

糖 1/4 茶匙，老抽 1 茶匙，水 3 茶匙，
粟粉半湯匙

1/4 tsp sugar, 1 tsp dark soy sauce, 3 tsps water,
1/2 tbsp cornflour

做法 Method

1. 各式甜椒洗淨，去籽及切條。
2. 牛扒洗淨及切片，用醃料醃片刻。
3. 燒熱油 1 茶匙，下牛扒略炒，備用。
4. 燒熱油，下甜椒炒香，加入牛扒炒至熟，注入
   調味料炒勻，即可食用。

1. Wash and remove the seed of bell pepper, then slice.
2. Wash and slice the sirloin steak. Marinate the sirloin
   for a while.
3. Heat 1 tsp of oil. Stir fry the sirloin briefly. Set aside.
4. Heat the oil. Stir fry the bell peppers. Add the sirloin
   and stir fry until done. Pour the seasoning and stir
   well. Serve.

薄牛肉金菇卷
Steak and Enokitake Mushroom Rolls

你可用煙肉代替牛肉片，不必加入醬汁，味道已足夠。

You may wrap the enokitake mushrooms in smoked bacon instead of beef. You don't need any additional sauce either.

材料
Ingredients

薄牛肉片 12 兩，金菇 1 包

450 g beef strips, 1 pack Enokitake mushrooms

醬汁
Sauce

蠔油 2 湯匙，水半碗，糖 1 茶匙

2 tbsps oyster sauce, 1/2 bowl water, 1 tsp sugar

獻汁
Cornflour Solution

粟粉 1 茶匙，水 2 湯匙

1 tsp cornflour, 2 tbsps water

做法
Method

1. 金菇及牛肉洗淨，瀝乾水分。將 1 片牛肉包上適量金菇，用牙籤串起固定。
2. 燒熱油，下金菇牛肉串煎香，盛起。
3. 煮滾醬汁，與獻汁拌勻，煮至濃稠，澆在牛肉卷上即可。

1. Rinse and drain with beef and Enokitake mushrooms. Put some Enokitake mushrooms on a strip of beef and roll it up. Use toothpick to fix it.

2. Heat the oil. Fry the steak rolls until fragrant. Set aside.

3. Bring the sauce to boil. Mix with the cornflour solution and cook until thicken. Pour the sauce on the steak rolls. Serve.

XO 醬 西 芹 牛 柳
Beef Slices with Celery in XO Sauce

牛柳 8 兩 (切片)，西芹 3 枝，XO醬 3 湯匙，
蒜茸 1 茶匙

300 g beef fillet (sliced), 3 stalks celery,
3 tbsps XO sauce, 1 tsp grated garlic

生抽 1 茶匙，糖 1/4 茶匙，粟粉半茶匙，
紹酒半茶匙

1 tsp light soy sauce, 1/4 tsp sugar,
1/2 tsp cornflour, 1/2 tsp Shaoxing wine

1. 牛柳用醃料醃 2 小時。
2. 西芹洗淨、切條。
3. 燒熱油 1 茶匙，爆香蒜茸，下西芹炒勻，下
   2 湯匙水，蓋上鑊蓋一會，將西芹炒至熟。
4. 燒熱油 1 湯匙，下牛柳炒至八成熟，加入西
   芹及 XO 醬炒勻，盛起，即可食用。

1. Marinate beef for 2 hours.
2. Wash and cut the celery into strips.
3. Heat 1 tsp of oil. Stir fry grated garlic. Add celery
   and stir well. Pour 2 tbsps of water. Cover with lid
   for a while. Then stir fry the celery until done.
4. Heat 1 tbsp of oil. Stir fry the beef until 80% done.
   Mix the celery and add XO sauce. Serve.

自煮飯堂 啖啖肉

作者	Author
蔡美娜	
策劃/編輯	Project Editor
Forms Kitchen 編輯委員會	Editorial Committee, Forms Kitchen
攝影	Photographer
	Mike Tsang
美術統籌及設計	Art Direction & Design
	Me
出版者	Publisher
	Forms Kitchen
香港鰂魚涌英皇道1065號	Room 1305, Eastern Centre, 1065 King's Road,
東達中心1305室	Quarry Bay, Hong Kong
電話	Tel　2564 7511
傳真	Fax　2565 5539
電郵	Email　info@wanlibk.com
網址	Web Site　http//www.formspub.com
	http//www.facebook.com/formspub
發行者	Distributor
香港聯合書刊物流有限公司	SUP Publishing Logistics (HK) Ltd.
香港新界大埔汀麗路36號	3/F., C&C Building, 36 Ting Lai Road,
中華商務印刷大廈3字樓	Tai Po, N.T., Hong Kong
電話	Tel　2150 2100
傳真	Fax　2407 3062
電郵	Email　info@suplogistics.com.hk
承印者	Printer
合群(中國)印刷包裝有限公司	Powerful (China) Printing & Packing Co., Ltd.
出版日期	Publishing Date
二〇〇九年四月第一次印刷	First print in April 2009
二〇一七年一月第五次印刷	Fifth print in January 2017

瀏覽網站

會員申請